Minerals and Rocks

Minerals and Rocks

A BEGINNER'S GUIDE

KIM TAIT

FIREFLY BOOKS

A Firefly Book

Published by Firefly Books Ltd. 2024
Copyright © 2024 Firefly Books Ltd.
Text copyright © 2024 Kim Tait
Photographs © as listed on page 56

All rights reserved. No part of this publication may be reproduced, stored in a retrieval system, or transmitted in any form or by any means, electronic, mechanical, photocopying, recording or otherwise, without the prior written permission of the Publisher.

First printing

Library of Congress Control Number: 2024934077

Library and Archives Canada Cataloguing in Publication
Title: Minerals and rocks : a beginner's guide / Kim Tait.
Names: Tait, Kimberly T. (Kimberly Terri), 1976- author.
Description: Includes index.
Identifiers: Canadiana 20240330595 | ISBN 9780228104957 (hardcover) | ISBN 9780228104964 (softcover)
Subjects: LCSH: Minerals—Handbooks, manuals, etc.—Juvenile literature. | LCSH: Rocks—Handbooks, manuals, etc.—Juvenile literature. | LCSH: Minerals—Identification—Handbooks, manuals, etc.— Juvenile literature. | LCSH: Rocks—Identification—Handbooks, manuals, etc.—Juvenile literature. | LCGFT: Handbooks and manuals.
Classification: LCC QE365.2 .T35 2024 | DDC j552—dc23

Published in the United States by	Published in Canada by
Firefly Books (U.S.) Inc.	Firefly Books Ltd.
P.O. Box 1338, Ellicott Station	50 Staples Avenue, Unit 1
Buffalo, New York 14205	Richmond Hill, Ontario L4B 0A7

Cover and interior design: Stacey Cho
Author photo (back cover): Courtesy of ROM (Royal Ontario Museum), Toronto, Canada. Paul Eekhoff ©ROM

Printed in China | E

Canada We acknowledge the financial support of the Government of Canada.

Dedication

This book is dedicated to my family, whose unwavering love and understanding have been my anchor. My husband, Sal; my five children, Emily, Matthew, Ethan, Owen and Zackery; and my dog, Pickles — all whom I love dearly. With each of you, I have been blessed with boundless love, immeasurable joy and endless inspiration.

Acknowledgments

Before delving into the pages of this book, I must express my heartfelt gratitude to those whose support, guidance and encouragement have made its creation possible. I am profoundly grateful for the unwavering support, encouragement and camaraderie of my colleagues at the Royal Ontario Museum, especially Katherine Dunnell and Veronica Di Cecco. I am deeply appreciative of the countless ways in which you have enriched and inspired me along the way. I am also indebted to my students and my research group, whose insightful questions, thirst for knowledge and eagerness to explore have enriched my career.

I extend my sincere thanks to the publishing team at Firefly who believed in this project and worked tirelessly to bring it to life. Your dedication, professionalism and passion for literature has made this collaboration a joyous and rewarding experience.

To the readers of this book, I am deeply grateful for your curiosity, engagement and support. It is my sincere hope that the content within this book resonates with you, provokes thought and continues you on your own journey of love for the natural world.

Lastly, to Earth itself — the ultimate teacher. I am endlessly grateful for the countless wonders you reveal and the lessons you impart. May we continue to explore, discover and protect the treasures you hold.

Table of Contents

Introduction 8
    The Elements and Our Earth 8
    What Is a Mineral? 9
    What Is a Rock? 9

Chapter 1: Minerals 11
    Identifying Minerals 11
- Shape 12
- Luster 14
- Hardness 17

    Feldspar 18
    Quartz 19
- Microcrystalline Quartz 21

    Mica 22
    Native Metals 26
    Critical Minerals 28
    Gemstones 30

Chapter 2: Rocks 33
    The Rock Cycle 34
    Igneous Rocks 36
    Sedimentary Rocks 38
- Clastic Rocks 39
- Chemical and Biochemical Rocks 40

    Metamorphic Rocks 42
- Foliated Metamorphic Rocks 42
- Non-foliated Metamorphic Rocks 44

Chapter 3: Becoming a Geologist 47
    Taking Field Notes 48
    Collecting Minerals and Rocks 49
    Testing Minerals and Rocks 50
    Resources for Young Geologists 52
- Books 52
- Websites 53
- Apps 53

Index 54
Photo Credits 56

Introduction

Minerals and rocks are all around us. They are in the ground we walk on, the buildings we live in and even the food we eat. Every mineral and rock on Earth has a story to tell. And if you learn how to understand these stories, a new and wondrous world will reveal itself to you.

The Elements and Our Earth

Elements* are the building blocks of everything. They are pure substances that are made from a single type of **atom**, and they cannot be broken down into smaller parts. Iron, oxygen, hydrogen, gold and helium are all common elements you might have heard of. There are over 90 naturally occurring elements. The Earth's outermost layer — the crust — contains nearly all of them.

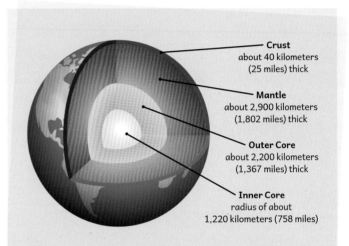

Crust
about 40 kilometers
(25 miles) thick

Mantle
about 2,900 kilometers
(1,802 miles) thick

Outer Core
about 2,200 kilometers
(1,367 miles) thick

Inner Core
radius of about
1,220 kilometers (758 miles)

Earth is a dynamic planet that has different layers. The outer layer is cool and solid. It is called the **crust**. This contains the minerals and rocks that we see every day and that we use in many ways. Below the crust lies a layer of hot, almost-solid rock called the **mantle**. At the center of the planet is the ball-shaped **core**, which is very hot and very dense. The core is made almost entirely of the metals iron and nickel.

*The words that appear in **bold** are key terms.

What Is a Mineral?

Minerals are materials that form naturally on our planet, on the moon and far beyond — like on other planets and asteroids. You can think of them like the geological building blocks of the Earth, and they combine in many ways to make different types of rocks and minerals. Minerals are usually solid and are formed by inorganic processes — meaning not by humans, other animals or plants.

Like everything else, minerals are made up of elements, but there are just eight elements that make up most minerals on Earth: iron, silicon, aluminum, calcium, sodium, potassium, magnesium and oxygen. Minerals have a regular repeating pattern of these elements. If they do not have a regular repeating pattern, then they are not considered minerals.

Sometimes when minerals have a lot of time and space to grow, they form beautiful shapes called **crystals**. A crystal is the natural shape that each mineral forms, and it is based on how the mineral's atoms lock together. **Gemstones**, or **gems**, are minerals that have been cut by humans into a particular shape for jewelry and other uses.

Smoky quartz, a variety of the mineral quartz

What Is a Rock?

A **rock** is a solid mixture of several different minerals and other geologic materials. These materials can include glass, other rocks and even fossils. There are three broad categories of rock, based on how they formed — **igneous**, **sedimentary** and **metamorphic** — and thousands of different types of rock that differ based on not just how they formed, but also what materials they're made of and the amount of each material.

The rest of this book will dive deeper into the amazing world beneath your feet. The following chapters will introduce you to some of Earth's most important and beautiful minerals and rocks and give you tools and tips to learn more about the rocks in your own backyard.

Gabbro, a type of igneous rock

Chapter 1
Minerals

There are over 5,500 minerals known to date — and we keep finding new ones. What elements a mineral is made of and how the elements come together, or **bond**, will determine its shape, color and physical properties, such as if it is magnetic or hard. There are about 200 common minerals, but some minerals are so unique that there is only one of them in the world. Take the mineral kyawthuite — only one known sample of this mineral has ever been found, in a remote region of Myanmar.

Minerals can form anywhere on Earth, from the surface to deep underground. How they form depends on the physical and chemical conditions of the environment where they come from. Some minerals develop slowly, over hundreds of thousands of years. Other minerals can crystalize over a few minutes, hours, days or years, like the small minerals that form in igneous rocks when hot magma cools quickly (see Chapter 2, page 36 to learn more about this process).

Identifying Minerals

There are many tools for identifying minerals, including streak plates and acids (see Chapter 3, page 50), but minerals also have certain physical qualities we can observe to help us pinpoint what they are, such as shape, luster and hardness.

Shape

When minerals have a lot of time to grow, the atoms form repeating patterns that translate to the shape of the larger crystal. When a mineral has a shape that helps identify it, this is called its **habit**.

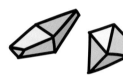

The mineral **fluorite** can form in any color of the rainbow; it even forms with stripes. Fluorite has many habits, but the most notable one is its octahedral shape, an eight-sided shape that looks like two pyramids have been stuck together at their bases.

The mineral group **tourmaline** typically forms in three-sided crystals. Tourmalines have the largest range of chemical compositions and more color banding within crystals than any other mineral group. Some tourmalines are pink in the center and green on the rim. These types are called **watermelon tourmalines**.

Garnets, another group of minerals, typically form as dodecahedrals, which are 12-sided shapes. Garnets can come in any color, from deep red (pyrope) to vibrant green (tsavorite). Some are a beautiful blue, some are colorless, and some even change color under different lights. Although garnet makes a beautiful gemstone, garnet is more often used in sandpaper, for sandblasting and for water filtration because it is hard and great for grinding and polishing.

Luster

Noting a crystal's habit is a good place to start when identifying a mineral. However, many minerals are **massive**, meaning they do not have a distinct crystal shape. That's why we must look at other qualities to help us identify minerals.

Luster describes how a mineral crystal's surface **reflects** light and how the light **refracts**, or bends, in transparent minerals.

Some minerals, like **hematite**, have a **metallic luster**. These minerals are opaque, meaning you cannot see through them, but the surface is shiny, like polished metal.

Minerals with a **non-metallic luster** can be subdivided into other types. **Diamond** has a very special luster, called **adamantine**. This means the mineral looks hard and shiny.

In contrast to diamonds, other minerals, like this **kaolinite**, can have a **dull** or **earthy luster**. This means the surface is non-reflective and looks quite plain.

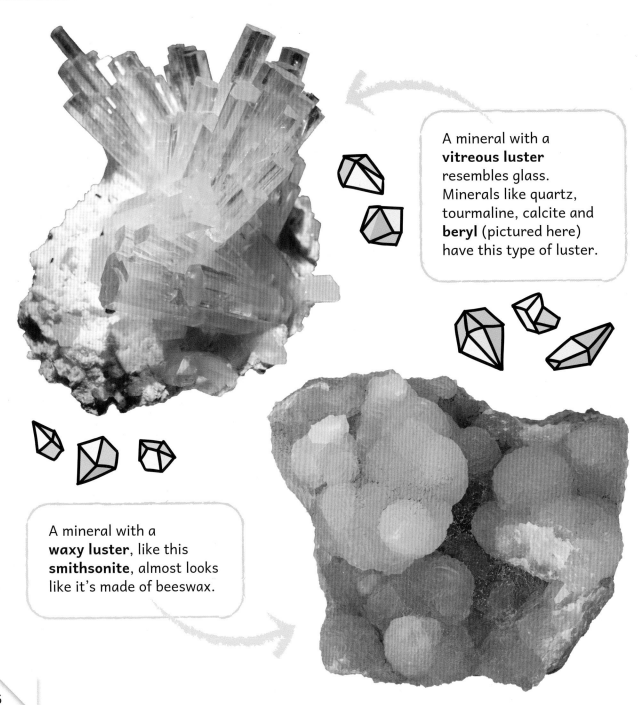

A mineral with a **vitreous luster** resembles glass. Minerals like quartz, tourmaline, calcite and **beryl** (pictured here) have this type of luster.

A mineral with a **waxy luster**, like this **smithsonite**, almost looks like it's made of beeswax.

Hardness

Testing **hardness** is another way to identify a mineral. Geologists use something called the **Mohs hardness scale** (and you can too; see Chapter 3, page 50). This scale helps measure how easily a mineral can be scratched by certain tools. The lower the number is on the scale, the softer the mineral is.

Hardness is especially important for gemstones. The gems that you put in jewelry need to be durable and hard. If a gem is too soft, the metal setting that holds it in place will wear the gem away and you would lose it. This is why hard gems like diamonds (which have a Mohs hardness of 10), rubies and sapphires (which both have a Mohs hardness of 9) are great for jewelry. Here's a beautiful **ruby** specimen. On the other hand, the mineral talc (which has a Mohs hardness of 1) is so soft, it is used in cosmetics to absorb moisture — not ideal for a ring or a necklace!

Now that you know the basics when it comes to describing minerals, let's dive into some of the most important mineral groups in the world.

One element that is very interesting is carbon. When the carbon atoms are bonded, or put together, in a certain way, the mineral can be very soft. This mineral is called **graphite** (which has a Mohs hardness of 1 to 2). Graphite is so soft that it is used in pencils. On the other hand, if the carbon atoms are put together a different way, it creates the hardest mineral, diamond. Minerals that have the same chemical formula, but different arrangements of atoms are called **polymorphs**.

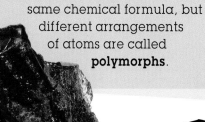

Feldspar

Feldspar is an important group of minerals that make up as much as 60 percent of the Earth's crust. This group has many different minerals in it, such as plagioclase, labradorite and orthoclase. Feldspar is commonly used in glassmaking, ceramics and sometimes paint, plastics and rubber. Most feldspars are reddish or pale pink and have a hardness of 6 on the Mohs scale, making them roughly as hard to scratch as glass. Granite, a type of rock that is used for countertops and building stones, is largely made up of feldspar.

Labradorite, a member of the feldspar group, is named after the region called Labrador in Canada, but it is found in other parts of the world. Labradorite exhibits a play of colors called **iridescence**. When you look at a sample, you can see blue, violet, green, yellow and even orange-red, and these colors change as you view the stone from different angles. This visual effect is so unique, it even has its own name — **labradorescence**.

Moonstone's milky, bluish hue is called **adularescence** (or schiller). The effect is caused by light **diffracting** (spreading) through the stone. Depending on how the light diffracts, the color can vary from white to blue and has a moonlight-like sheen — hence the stone's name.

Quartz

Quartz is the second most common mineral in the Earth's crust after feldspar. When quartz is made up of just silicon and oxygen, it is colorless. However, if it contains just a small number of other elements, its color can change to almost any in the rainbow. Quartz has a Mohs hardness of 7.

Quartz forms in a lot of different shapes and sizes, from large, perfectly shaped, six-sided crystals to smoothed-out pebbles in streams. When quartz is clear and colorless and forms in a perfect crystal shape, it is called **rock crystal**, shown here.

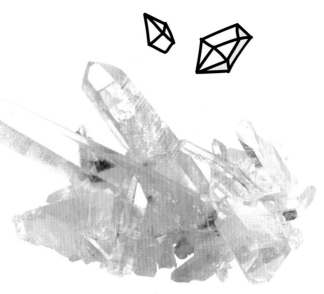

Most of the **sand** at the beach is quartz, which doesn't have much shape due to a process called **weathering**. Weathering is when a mineral's shape is rounded due to wind, water, cold temperatures and heat over a long period of time.

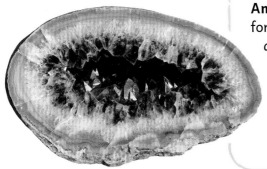

The most common non-clear color of quartz is purple. **Amethyst**, the purple version of quartz, mainly forms inside volcanic rock. If gas gets trapped inside a volcanic rock and the amethyst crystals inside have a long time to grow, a **geode** is formed. The most famous amethyst geodes come from Brazil. The outside of the geode looks brown and plain, but on the inside are some of the most beautiful purple crystals.

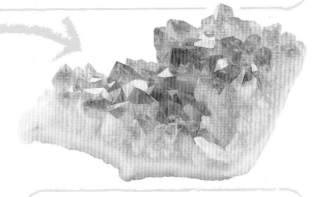

Citrine is the pale yellow to brownish-orange variety of quartz. Quartz with this natural color is quite rare, but it was discovered that if you heat up purple amethyst, it will turn a beautiful yellow. Much of the citrine that is sold now was purple amethyst at one time.

Sometimes you will see minerals that are half-amethyst and half-citrine, called **ametrine** (which is a combination of the two mineral names). There is one place in the world where ametrine forms naturally — in eastern Bolivia. Most ametrine is made in a lab by heating up half an amethyst crystal.

Smoky quartz is the brown to black version of quartz. Its color is due to natural radiation in the rock surrounding it.

Microcrystalline Quartz

Chalcedony is the term used for several varieties of microcrystalline quartz (quartz that is made up of small crystals). This type of quartz comes in a bunch of different colors and has many different names depending on the pattern or color the small crystals form. Some examples are agate, flint, chert, jasper, onyx, bloodstone and carnelian.

Agate, a type of chalcedony, is characterized by patterns of colored bands inside typically volcanic rocks. A particularly striking example of agate is called **tiger's eye**. It is very common for minerals and rocks to get their names from the things in nature that they look like. This is a great example. It has alternating bands of brown, bright gold to red and dark brown that mimic the eyes of a tiger.

Mica

Mica is a group of silicate minerals that form in flat layers, like pages in a book. Individual mica crystals can be split into very thin, flexible plates. Mica is used in cosmetics, such as nail polish, lip gloss and eye shadow. It adds "shimmer" to these products because when light shines on the small plates, it creates a glittery effect. Mica is also used as a filler in cement and asphalt and as an insulation material in electrical cables.

Mica minerals come in a variety of colors. **Muscovite** is the clear-to-light-brown version of this group. **Phlogopite** is light brown to brown. **Lepidolite** is the purple version of this group, and **biotite** (or, more correctly, **annite**) is the black version.

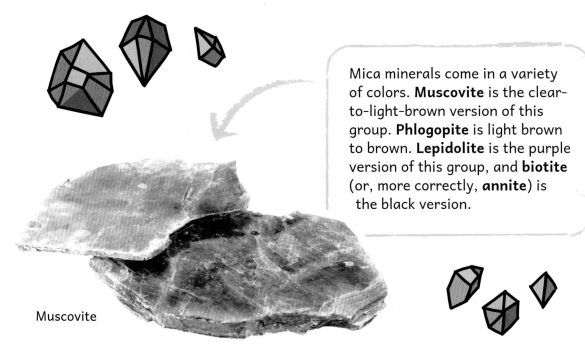

Muscovite

Sometimes the name of the rock identifies what minerals are present in the rock. For example, a **garnet mica schist** is a fine-grained, foliated (thin sheets of minerals that are repeated) metamorphic rock with large garnet minerals in it and layers of mica. (See Chapter 2, page 43 to learn more about schist rocks.)

Lepidolite

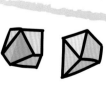

Biotite, or annite

MINERALS THAT GROW FROM WATER

In some places where there is water, there are dissolved minerals in that water. As the water **evaporates**, or dries up, it leaves the minerals behind, and these eventually turn into solid minerals. For example, as saltwater dries up, the salt content solidifies and makes a mineral called **halite**, which we use as salt on our food. Halite, which is made up of sodium and chlorine, forms in perfect cubes. Even when it is crushed into tiny pieces, it breaks apart into perfect cubes. Have a close look next time you have French fries with salt on them.

GYPSUM

Another mineral that forms from evaporated water is **gypsum**, which is a calcium sulfate dihydrate ("dihydrate" means it contains two water molecules). Gypsum is used in building materials, such as drywall (also known as gyprock or plasterboard). It is also used in blackboard chalk, fertilizers, some paints, toothpastes and shampoos.

Gypsum is soft (it has a Mohs hardness of 2), so sometimes it is polished into spheres, hearts and other shapes. Sometimes gypsum mixes with sand to create interesting formations called **desert roses**.

The crystals of gypsum are sometimes referred to as **selenite**. Selenite is usually white and forms in long rods. In fact, giant selenite crystals have been found in the caves of the Naica Mine in Mexico. These crystals are up to 11 meters (36 feet) long. The crystals grow to such heights because the cave has hot, stable temperatures of 58°C (136.4°F) and it is filled with mineral-rich water that feeds the crystals' growth.

Desert roses made of gypsum.

Giant selenite crystals in the Naica Mine, Mexico. See the person in the bottom right of the photo for an idea of how big these crystals are.

CALCITE

Calcite is a very common mineral that is found in most places on Earth. Pure calcite is colorless or white, but small amounts of other elements can turn it red, yellow, green, pink or blue — pretty much any color. When calcite is colorless, transparent and in its natural form of a rhombus, it is called **Iceland spar** or **optical calcite**.

One of the best ways to identify calcite is with a weak acid. When you place an acid such as white vinegar on calcite, it will bubble. This is because when the calcium carbonate of the calcite is combined with an acetic acid (vinegar), carbon dioxide is released.

Native Metals

Most of the elements on the periodic table are considered "metals." Gold, tin, iron and nickel are all examples of metallic elements. When these elements are found in their pure forms, they are called **native metals**. Most native metals will mix with other metals to form **alloys** or will mix with oxygen to form other minerals. Only gold, silver, copper and metals from the platinum group of elements occur on Earth in large amounts.

Don't be fooled, though. There is another mineral that looks like gold but isn't. It is called **pyrite**, also known as "fool's gold." Pyrite has a brassier color than gold and a different habit (or shape).

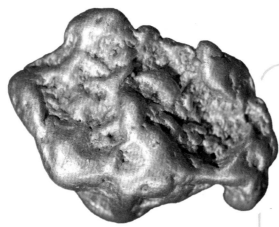

Gold might be one of the most famous minerals. It is a soft and heavy mineral with exceptional properties. For example, gold can be hammered without breaking and is easily stretched. It's a good conductor, meaning heat and electricity easily flow through it. Gold is used a lot in jewelry, and great achievements are usually celebrated with gold, like medals, trophies and other awards. Since gold is so valuable, when it is found, it can create something called a "gold rush" to get it.

Silver has a bright, metallic luster and a white color when untarnished. Tarnishing is when silver reacts with sulfur in the air to make silver sulfide, the dark spots on silverware or silver jewelry that need to be polished off.

Silver is usually found with quartz, gold, copper and minerals that contain silver. Silver likes to bond to non-metallic elements to make new minerals or combine with other metals to make alloys. A common alloy of silver is called **electrum**, which is when there is at least 20 percent silver mixed with gold.

Copper is a reddish-brown mineral that turns green when it's exposed to air for a long time. Copper has excellent electrical conductivity, meaning it can easily transfer electricity from one place to another. This is why we use copper wires for electricity.

Critical Minerals

Critical minerals are minerals such as copper, nickel, cobalt and lithium that are essential in the production of clean-energy technologies like solar panels, wind turbines, semiconductors and electric vehicles.

As a society, we have relied heavily on fossil fuels to power our homes, cars and buildings. But burning fossil fuels causes pollution and the buildup of greenhouse gas emissions, which is warming our planet. To stop using fossil fuels, we need to start transitioning to green energy. This will require sourcing and mining specific elements from our planet. Each country has a different list of critical minerals, but they do share a few things in common:

- There are few or no substitutes,
- They are strategic (essential to economic and/or national security) and somewhat limited in the amount of each mineral there is, and
- Sometimes they are concentrated in a certain place on the planet.

Clean-energy technologies, like wind turbines, will help us transition away from using fossil fuels. Certain minerals will be important in making these technologies more widespread.

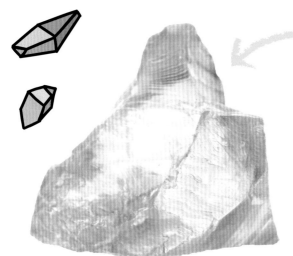

Lithium is a critical mineral, but it does not form as a native metal alone. Instead, it is found in brines (water that has a high amount of salt and other minerals) and in another mineral called **spodumene**. Lithium is used in electric vehicle batteries and is sometimes thought to be the next oil because it will become so valuable to carmakers.

Other elements that are needed to replace fossil fuels include aluminum, molybdenum, copper, **indium** and cobalt. Understanding where these elements are found, what types of rocks they form in and how to get them out of the ground (or **extract** them) will help us transition away from fossil fuels and achieve "net zero" — which means cutting greenhouse gas emissions to as close to zero as possible. To get there, we will need a lot more geologists. See Chapter 3 to find out how you can get started!

Indium is used to produce indium tin oxide, which is an important part of solar panels.

Gemstones

A gem or a gemstone is a mineral that usually has a very nice color and can be cut, or **faceted**, into a shape. Most rough (or natural) gems are crystals. A mineral's natural crystal shape can be further cut to enhance the color and brilliance. We cut **transparent** gems (ones we can see through) to make light travel through them and back to our eyes. This is what makes them sparkle.

Every time you cut a gem, you lose a bit of material. Since gems are usually sold by their size, gem cutters try to cut away as little as possible. For example, a diamond crystal looks like two pyramids stuck together at the bases. A diamond cutter usually tries to cut two gemstones from the one crystal, with each base of the pyramid now becoming the top of the gem.

Some gems are **opaque**, meaning no light can pass through them. These stones can be cut so that they are flat, or sometimes they are rounded into beads or carved into shapes. Sometimes these stones are cut in a certain way to show off a pattern in the mineral.

A rough diamond crystal (on the left) and a diamond that has been faceted (on the right)

Prices for gems can range from just a few dollars to millions of dollars, depending on how rare the gem is, its color, its shape and its size. Some gems are so precious and rare they are kept in museums, held by royalty or owned by celebrities.

BIRTHSTONES

Birthstones are gemstones that represent the month a person was born in. Some researchers trace the origins of birthstones to religious traditions around the world.

Modern birthstones were standardized in 1912, continuing the custom of giving someone a beautiful gemstone to represent when they were born.

January: Garnet

February: Amethyst

March: Aquamarine

April: Diamond

May: Emerald

June: Pearl

July: Ruby

August: Peridot

September: Sapphire

October: Opal

November: Topaz

December: Turquoise

Chapter 2
Rocks

As you may recall, a rock is a solid mixture of several different minerals and other geologic materials. These materials can include glass, other rocks and fossils. There are three types of rocks on our planet: **igneous**, **sedimentary** and **metamorphic**. The rock cycle is the name for the processes that change one rock type into another rock type.

The Rock Cycle

Unlike other planets in our solar system, our dynamic planet has four spheres that interact with each other: the geosphere, the hydrosphere, the atmosphere and the biosphere. These four spheres interact constantly and rely on each other for balance.

The **geosphere** (sometimes called the lithosphere) is all the rock on Earth. The geosphere starts on the surface of the Earth and goes all the way to the core of the planet. Humans rely on the geosphere to provide natural resources to heat our homes and for soil to grow our food.

The **hydrosphere** is all the water on Earth. It includes oceans, rivers, lakes, streams and water frozen in glaciers. We rely on water to survive, not only to drink but also to grow food. All other animals and plants depend on water as well.

The **atmosphere**, which is more commonly called the air, is all the gases surrounding Earth. All planets have an atmosphere, but as far as we know, Earth is the only planet with the correct combination of gases to support life.

The **biosphere** is made up of all living things on Earth and includes insects, fishes, birds, plants and humans.

The Earth's geosphere interacts with the other spheres in unique ways compared to other planets. Because of this, our geosphere is constantly changing. Here are some of the processes that change rocks to make new rocks:

1. **Weathering and erosion:** Rocks can be broken down by the hydrosphere, the atmosphere and even sometimes the biosphere.

2. **Melting:** With increased pressure and temperature, solid rocks can turn into liquid or semi-liquid rocks.

3. **Lithification:** With some pressure and time, loose particles can be turned into new rock.

4. **Cooling:** When melted rocks are brought to the surface and interact with the hydrosphere and atmosphere, they can turn into new rocks.

Think of our planet as the great rock recycler!

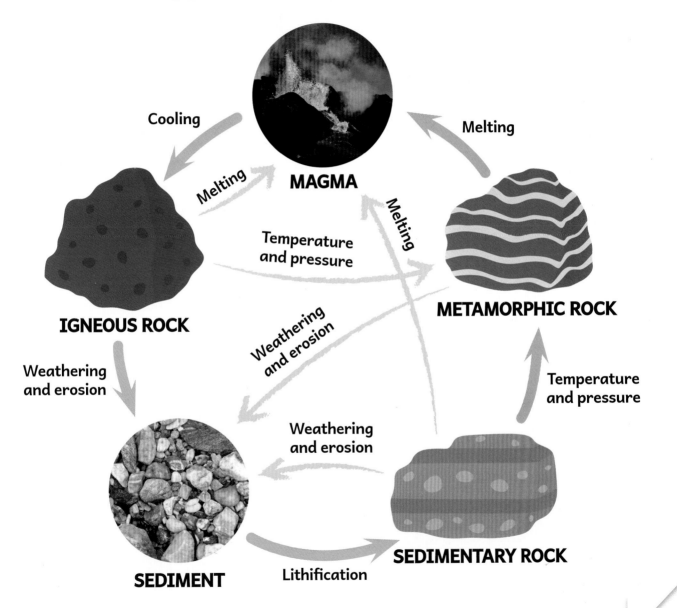

Igneous Rocks

Magma is very hot liquid or semi-liquid rock below the Earth's crust. **Igneous rocks** are formed by cooled magma, either on the surface of the Earth or in the Earth's crust. The word "igneous" comes from the Latin word *ignis*, meaning "fire."

Intrusive rocks are igneous rocks that form when magma is trapped deep underground. These types of rocks cool and solidify very slowly — over thousands, sometimes even millions of years. The slow cooling of the rock means that mineral crystals in the rock have a long time to grow. The texture of intrusive rocks tends to be coarse with visible mineral crystals, like those in granite, gabbro and pegmatite.

A basaltic lava flow in Hawaii, USA.

Granite is a coarse igneous rock. It is made up of about 30 percent quartz (the white grains), about 65 percent feldspar (the pink grains) and sometimes black mica or amphibole minerals. Granite is a hard rock and is used all over the world for everything from kitchen countertops to monuments, buildings and road paving.

Extrusive rocks are igneous rocks that form on or near the surface of the Earth. In certain places, hot magma comes from inside the Earth to the surface from a **volcano** or fissure. When magma gets to the surface, it cools quickly, especially when it interacts with cold ocean waters. The quick cooling of the rock means that the mineral crystals in the rock don't have much time to grow, resulting in rock that is finer-grained, like basalt.

Basalt is a fine-grained, dark-gray or black igneous rock. It makes up about 90 percent of all the lava rock on Earth. The main minerals that form in basalt are feldspar, pyroxene, olivine and iron oxide.

Sometimes extrusive igneous rocks cool so fast, they don't even have time to form into rock at all. Instead, they become a **volcanic glass**. **Obsidian** is a naturally occurring, hard, brittle volcanic glass. Glass has elements like a mineral does, but it does not have a regular repeating atomic structure. Without this, these glassy rocks break in irregular ways, called **conchoidal fractures**.

When extrusive igneous rocks have gases in them, the gases can get trapped in the lava and the lava cools around them. Depending on how many gassy holes there are in the rocks, this type of rock is called **scoria** (which is denser, pictured here) or **pumice** (which is lighter). Pumice rock is so full of holes, or voids, it can float. It is quite hard and light, so it is sometimes used for cleaning ceramics and rubbing dead skin off your feet.

Sedimentary Rocks

Sedimentary rocks are made by **erosion**, or the breaking down of solid rock by liquid water, wind and ice. Small pieces of rock break down to even smaller pieces called **sediment**. The sediment can travel around by wind and water, and mix with pieces of plants and animals too. Once it settles, layers of sediment build up over thousands of years. As the layers of sediment build up — just like you piling more and more blankets on top of yourself — they are compressed, or squashed, and become bonded together. If the pieces of sediment are too big, sometimes the spaces around the bigger pieces get filled in with dissolved minerals, which make a kind of cement when they harden.

Sedimentary rocks can also form when the shells of sea creatures, which are made of minerals, fall to the bottom of the ocean after the creatures die. Sometimes, when a larger animal like a fish, bird or dinosaur dies and is surrounded by sediment, the hard parts of their bodies — bones, teeth and shells — can be preserved. Even plants, under the right conditions, can be preserved in sediment.

This river system in southern Iceland transports sediment created by a glacier.

Sedimentary rocks can be grouped according to the way they are created and the size of the pieces of rock they contain. Two of the main groups of sedimentary rocks are clastic rocks and chemical and biochemical rocks.

Clastic Rocks

Clastic sedimentary rocks are made up of pieces (called **clasts**) of existing rocks. Often the names of the rocks are based on the size of the grains they contain. Clay is the smallest grain, followed by silt, sand and pebbles, which are larger than 2 millimeters.

Mudstones are rocks made up of particles of clay and silt. They form where clay and silt have settled in calm water, such as in lakes, lagoons or the deep sea.

Sandstones are rocks made up of pieces of sand (0.05-to-2-millimeter grains) that are cemented together. Sandstones are formed from sediment deposited from rivers, from the sea or by wind, so there are many types of sandstones.

Conglomerates are formed when large pebbles are glued together with small particles of sand and clay. The rock color can vary depending on the type of pebbles in the rock. They are usually formed where there is fast-moving water, like rivers and beaches.

Chemical and Biochemical Rocks

Chemical sedimentary rocks form when a liquid like water dissolves some of the minerals in existing rocks and carries them away. The minerals are deposited elsewhere and build up when the water evaporates. **Biochemical sedimentary rocks** form when living organisms that use minerals to build their bodies die and pile up, eventually compressing and turning into rock.

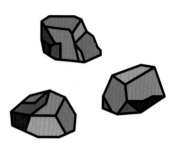

Limestone is a sedimentary rock that is made up of calcite and aragonite, two different mineral forms of calcium carbonate. The calcium carbonate comes from the bodies of organisms that lived long ago. **Chalk** is a type of limestone. It is mostly formed from the tiny calcium carbonate scales of phytoplankton, which are microscopic marine algae. In the south of England, there are thick deposits of chalk called the **White Cliffs of Dover**, shown here. The chalk that you use in the classroom with the blackboard or on the sidewalk used to be made of chalk, but it is now mostly made of the mineral gypsum (see page 24).

Freshly cut blocks of peat stacked in rows to dry in the sun.

Coal is a type of sedimentary rock that we burn to produce energy. The coal we use formed hundreds of millions of years ago. At that time, our planet had swamps with giant plants. As the plants died, they sank to the bottom of the water, and the plant debris that didn't decay completely created something called **peat**. The peat was buried by sediment, and over a long period of time the heat and pressure of the Earth turned the peat into coal. Coal is a fossil fuel, and it is the only solid fossil fuel.

Not all types of coal are the same. **Anthracite** is the densest coal, and it is shiny and black. **Bituminous** coal is also black, but softer. **Lignite**, the softest form of coal, is dull and brown.

Anthracite

Bituminous coal

Lignite

Metamorphic Rocks

The word "metamorphic" comes from the Greek words *meta*, meaning "change," and *morphe*, meaning "form." So **metamorphic rocks** are rocks that change their form. These rocks start out as one type of rock deep in the Earth, and with pressure, heat and time, they gradually change into a new type of rock.

Metamorphic rocks can be formed from sedimentary rocks, igneous rocks or another kind of metamorphic rock. For example, limestone is a sedimentary rock that can be metamorphosed into a rock called marble.

Foliated Metamorphic Rocks

When rocks are heated and squeezed, some of the preexisting minerals are flattened and spread out. Sometimes they even create patterns of stripes and layers, called **foliation**. There are three types of foliations in metamorphic rocks: slate, schist and gneiss.

The limestone shown at the top is changed to marble like this when intense heat and pressure recrystallize the mineral grains.

Slate has thin layers with small mineral grains. In fact, it has the smallest grains of the three types of foliated rocks discussed here. It metamorphosed from shale, a sedimentary rock that is composed of quartz, mica, iron oxides and iron sulfides. This rock can break along layers to make a very flat surface — so flat that this type of rock was used to write on before paper was readily available.

Schist has thin layers with large mineral grains. Schists contain lots of flattened minerals that have been drawn out due to heat and pressure, such as mica. These flat minerals are usually mixed with quartz and feldspar. Schists are often named after the important minerals in them. For example, **mica schists** contain mica, garnet schists contain garnet, tourmaline schists contain tourmaline, and so on.

Gneiss (pronounced "nice") is a sedimentary rock (like sandstone) or an igneous rock (like granite) that has been exposed to extreme temperatures and pressures. When this happens, almost no traces of the original rock remain. The mineral grains stretch out to make alternating layers of light and dark bands.

Non-foliated Metamorphic Rocks

Non-foliated metamorphic rocks lack the parallel arrangement of minerals that the foliated rocks have. Instead, they have the same properties in all directions. In these rocks, minerals from the original rock can start off small, but with heat and pressure the minerals can grow to be bigger. The three most common types of non-foliated metamorphic rocks are quartzite, marble and hornfels.

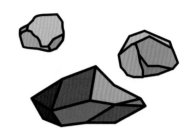

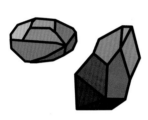

Quartzite forms by the metamorphism of sandstone, and most sandstones are mainly made up of the mineral quartz. Quartzite is usually white or gray. The grains of quartz are usually around a millimeter in size and form an extremely hard rock.

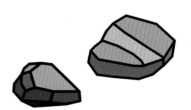

Marble forms by the metamorphism of limestone or a rock called dolostone. Limestone is made up mostly of calcite, and dolostone is made up mostly of dolomite. Marble is typically white, but if other minerals are present, it can be other light colors, like gray, pink and yellow. Since it has no foliation, this is a great rock for building stones and statues.

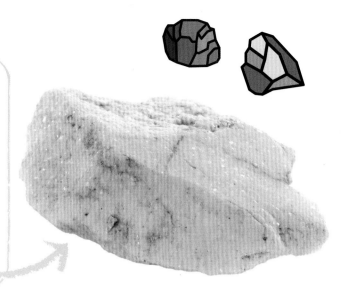

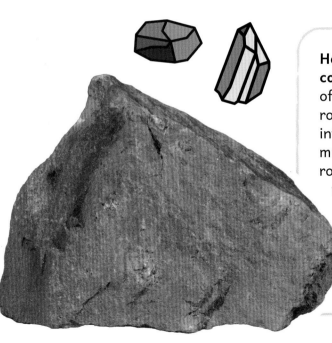

Hornfels forms by a process called **contact metamorphism**, when a body of magma quickly heats the surrounding rock and transforms it (with no pressure introduced). Hornfels comes from mudstone, shale or other clay-rich rocks. It lacks foliation and is dark, and most of the grains in hornfels are too small to see with the naked eye. The most common hornfels is **biotite hornfels**. It is dark brown to black, with sparkles from the small crystals of mica in the rock.

Chapter 3

Becoming a Geologist

Now that you know much more about minerals and rocks, you might dream of one day becoming a **geologist**. A geologist is a type of scientist that studies the Earth and what it is made of. This branch of science is called **geology**, and you can become an expert in a part of geology. For example, people who study minerals are called **mineralogists**, or people who study gems are called **gemologists**.

A great starting point for becoming a young geologist is to learn more about the rocks and minerals around you. You might have a rock collection or live in an area with interesting geologic formations. You could take field notes, collect specimens and test some of the rocks you collect.

Taking Field Notes

As you go out into your area looking for different types of rocks and minerals, bring a pencil and a notebook to take field notes about what you see around you. Here are a few things it helps to record for every rock and mineral you find:

- **The location** – Jot down the name and a description of your location. Bonus points if you record the GPS coordinates (certain phone apps can help you with this). Recording the location is important, especially if you want to return later.

- **A sketch of the rock or mineral** – Sketching can help you carefully and closely observe the features of the rock or mineral. You can also take a photo with a camera phone and compare it to your sketch later.

- **A description** – Write down any observations you can make about the rocks and minerals you find. For instance, how big are they? Are they heavy? Shiny? What color are they? What textures do they have?

A geology student takes notes on a field trip.

Collecting Minerals and Rocks

As you become more confident, you might bring a small hammer, safety goggles and sturdy shoes to break open rocks or collect samples. If the rock or mineral is on private land, be sure to ask permission before entering the area and breaking or taking any rocks. In certain areas, you are prohibited from taking any rocks away from a site, especially areas of cultural or natural significance, so make sure you know the rules before you take a sample.

When you bring your minerals and rocks home, you will need to label them. You can do this by placing a unique number on your rock with correction fluid (like Wite-Out®) or by putting your rock or mineral inside a numbered box. Make sure you update your field notes with the numbers. This is what museums do: All samples have specimen numbers to help **curators**, the people that take care of the museum collections, know where the samples are from and other details. That way, when someone wants to study the mineral or rock later, they have all the information about the sample at hand.

The Geomineral Museum in Madrid, Spain.

Testing Minerals and Rocks

There are several ways to test and identify minerals and rocks at home. Some minerals can be identified by color alone, but color can be tricky, as some minerals can form in any color of the rainbow. You can use a magnifying glass to see smaller features, as well as a magnet to test whether a mineral is magnetic (like the aptly named **magnetite**).

In Chapter 1, you were introduced to the Mohs hardness scale, which measures a mineral's scratch resistance. The hardness tests were designed to include a small group of tools that someone might have on hand when they're in the field. You are likely to have these same tools at home, so you can try it out on your rock or mineral specimens.

Each tool has an assigned hardness. For example, a fingernail has a Mohs hardness of 2.5. This means any minerals that are softer than this can be scratched with your fingernail. A copper penny has a Mohs hardness of 3.5. A metal knife has a Mohs hardness of 5.5, and a steel nail has a hardness of 6.5. Since diamond has a Mohs hardness of 10, it cannot be scratched by your fingernail or even a steel nail.

Mineral Name	Scale Number	Common Object
Diamond	10	
Corundum	9	Masonry Drill Bit (8.5)
Topaz	8	
Quartz	7	Steel Nail (6.5)
Orthoclase	6	Knife/Glass Plate (5.5)
Apatite	5	
Fluorite	4	Copper Penny (3.5)
Calcite	3	
Gypsum	2	Fingernail (2.5)
Talc	1	

Increasing Hardness →

Hematite (on the left) leaves a characteristic red streak, while pyrite (on the right) leaves a black streak.

A **streak plate** is a small, unglazed porcelain surface that minerals can be dragged on to help identify them. Some minerals, like **hematite**, have a characteristic streak color. Hematite is typically steel gray to black, but when you do a streak test on hematite, it always leaves a red streak. This distinguishes it from different minerals that are also steel gray to black.

As touched on in Chapter 1, weak acids like vinegar can help distinguish calcite from other white minerals. When you drop a bit of acid on a piece of calcite or limestone, small bubbles form on the surface where the acid was dropped, as shown here.

Be sure to record your observations and test results in your field notes. You can use mineral identification guides or online resources to narrow down the type of rock or mineral you have. Once you know what it is, you can learn all about its features, formation and much more.

Resources for Young Geologists

Here are some great resources for learning more about rocks and minerals.

Books:

- *Geology for Kids: A Junior Scientist's Guide to Rocks, Minerals, and the Earth Beneath Our Feet* by Meghan Vestal

- *Planet Earth for Kids: A Junior Scientist's Guide to Water, Air, and Life in Our Ecosphere* by Stacy W. Kish

- *National Geographic Kids Everything Rocks and Minerals* by Steve Tomecek

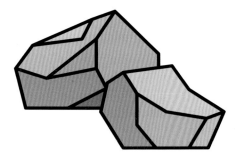

Websites:

- Natural Geographic Kids, Geology 101: **kids.nationalgeographic.com/science/article/geology-101**

- Smithsonian National Museum of Natural History, Earth Science Resources: **naturalhistory.si.edu/education/teaching-resources/earth-science**

- American Museum of Natural History, If Rocks Could Talk — Three Types of Rocks: **amnh.org/explore/ology/earth/if-rocks-could-talk2/three-types-of-rock**

- United States Geological Survey, Geology: **usgs.gov/science/faqs/geology**

- Geology.com, What Are Minerals?: **geology.com/minerals**

- Mineralogy4Kids, Welcome to MSA's Rockin' Internet Site: **min4kids.org**

- National Park Service, Rocks and Minerals: **nps.gov/subjects/geology/rocks-and-minerals.htm**

Apps:

- **Rockd** – an app to learn about, explore and document your geologic environment (free, available for iPhone and Android)

- **Smart Geology - Mineral Guide** – an app to help identify minerals and learn more about mineral classification, geologic time scales and geology terms (free, available for Android)

- **Geology by Kids Discover** – an app designed for kids to learn more about rocks, minerals, fossils and geology, with interactive 3D models, videos, photos, games and more (paid, available for iPhone)

Index

Page numbers in *italic* mean photos or charts.

acid, 25, 51
adamantine luster, 15
adularescence, 18, *18*
agate, 21, *21*
air, 34
alloys, 26–7
amethyst, 20, *20*, *31*
ametrine, 20, *20*
annite, 22, *23*
anthracite, 41, *41*
apatite, *50*
apps for geology, 53
aquamarine, *31*
aragonite, 40
atmosphere, 34
atoms, 8, 9, 12, 17

basalt, 37, *37*
beryl, 16, *16*
biosphere, 34
biotite, 22, *23*
biotite hornfels, 45
birthstones, 31, *31*
bituminous, 41, *41*
bones, 38
books about geology, 52
brines, 29

calcite, 25, *25*, 40, 45, *50*, 51
calcium carbonate, 25, 40
carbon, 17, *17*
caves, 24, *25*
chalcedony, 21
chalk, 40, *40*

chemical/biochemical rocks, 38, 40–1, *40–1*
citrine, 20, *20*
clastic rocks, 38–9, *39*
clay, 39, 45
clean-energy technologies, 28–9
coal, *41*
colors, 50
conchoidal fractures, 37
conglomerates, 39, *39*
contact metamorphism, 45, *45*
cooling, 34, *35*, 36–7
copper, 27, *27*
core of earth, 8, *8*
corundum, *50*
cosmetics, 22
critical minerals, 28–9, *29*
crust, 8, *8*
crystals
  defined, 9
  of gems, 30
  of gypsum, 24, *25*
  in intrusive rocks, 36
  of mica, 22
  in microcrystalline quartz, 21
  of quartz, 19, *19*
  rock crystal, 19, *19*
  selenite, 24, *25*
  and shapes, 12
curators, 49

desert roses, 24, *24*
diamonds, 15, *15*, 17, 30, *30*, *31*, *50*

diffracting light, 18
dodecahedral shapes, 13, *13*
dolomite, 45
dolostone, 45

earth layers, 8, *8*, 34
earth spheres, 34
electrical conductivity, 27
electrum, 27
elements, 8–9
emeralds, *31*
erosion, 34, *35*, 38
evaporation, 24
extraction, 29
extrusive rocks, 37

facets, 30
feldspar, 18, *18*, 36, 37, 43
field notes, 48, 51
fluorite, 12, *12*, 50
foliated metamorphic rocks, 42–3, *43*
fool's gold, 26, *26*
fossil fuels, 28–9, 41

gabbro, 9
garnet mica schist, 23, *23*
garnets, 13, *13*, *31*, 43
gases, 37
gemologists, 47
gems/gemstones, 9, 17, 30–1, *30–1*
geodes, 20
geology/geologists
  about, 47

apps for, 53
books about, 52
collecting, 49
field notes, 48, 51
testing, 25, 50, *50*
websites about, 53
geosphere, 34
glass, 9, 37, *37*
global warming, 28–9
gneiss, 42–3, *43*
gold, 26, *26*
GPS coordinates, 48
granite, 18, 36, *36*
graphite, 17, *17*
greenhouse gas, 28–9
gypsum, 24, *24*, 40, *50*

habit, 12, 14
halite, 24
hardness, 17. *See also* Mohs hardness scale
hematite, 14, *14*, 51, *51*
hornfels, 45
hydrosphere, 34

Iceland spar, 25
igneous rocks, 9, *9*, *35*, 36–7, *36–7*, 42, 43
indium, 29, *29*
inner core, 8, *8*
intrusive rocks, 36
iridescence, 18

kaolinite, *15*
kyawthuite, 11

labeling, 49
labradorescence, 18
labradorite, 18, *18*
lava, *36*, 37
layers of earth, 8, *8*
lepidolite, 22, *23*
lignite, 41, *41*
limestone, 40, *40*, 42, *42*, 45, 51
lithification, 34, *35*
lithium, 29, *29*
lithosphere, 34
luster, 14–16

magma, *35*, 36–7, *36*, 45
magnets, 50
mantle, 8, *8*
marble, 42, *42*, 45, *45*
massive minerals, 14
melting, 34, *35*
metallic luster, 14
metamorphic rocks, 9, *35*, 42–5, *42–5*
mica, 22–3, *22–3*, 36, 43, 45
mica schists, 43
microcrystalline quartz, 21, *21*
mineralogists, 47
minerals
  critical minerals, 28–9, *29*
  defined, 9
  feldspar, 18, *18*, 36, 37, 43
  gemstones, 30–1, *30–1*
  habit of, 12, 14
  hardness of, 17
  identifying, 11–17
  luster of, 14–16
  mica, 22–4, *22–4*, 36, 43, 45

microcrystalline quartz, 21, *21*
native metals, 26–8, *26–7*
number of, 11
quartz, 9, 19–21, *19–21*, 36, 43–4, *50*
shapes of, 12–13
water grown, 24–5, *24–5*
See also Mohs hardness scale
Mohs hardness scale, *50*
  about, 17
  of diamonds, 17
  of feldspar, 18
  of graphite, 17
  of gypsum, 24
  of quartz, 19
  of sapphires, 17
  of talc, 17
  tests and, 50
moonstones, 18
mudstones, 39, *39*, 45
muscovite, 22, *22*
museums, 49, *49*

native metals, 26–8, *26–8*
non-foliated metamorphic rocks, 44–5, *44–5*
non-metallic luster, 15

obsidian, 37, *37*
octahedral shapes, 12, *12*
opals, *31*
opaqueness, 14, 30
optical calcite, 25
orthoclase, 18, *50*
outer core, 8, *8*

pearls, *31*
peat, 41, *41*
pencils, 17
peridot, *31*
phlogopite, 22
polymorphs, 17
pumice, 37, *37*
pyrite, 26, *26*, 51, *51*

quartz, 9, 19–21, *19–21*, 36, 43–4, *50*
quartzite, 44, *44*

radiation, 21
reflection, 14
refracting, 14
resources, 52–3
rock crystal, 19
rocks
  categories of, 9, *35*
  chemical/biochemical rocks, 38, 40–1, *40–1*
  clastic rocks, 38–9, *39*
  defined, 9
  foliated metamorphic rocks, 42–3, *42–3*
  igneous rocks, 9, *35*, 36–7, *36–7*, 42, 43
  metamorphic rocks, 9, *35*, 42–5, *42–5*
  non-foliated metamorphic rocks, 44–5, *44–5*
  rock cycle, 33–4, *35*
  sedimentary rocks, 9, *35*, 38–41, *39–41*
rubies, 17, *17*, *31*

saltwater, 24
sand, 19, 39
sandstones, 39, *39*, 44
sapphires, 17, *31*

schist, 42–3, *43*
scoria, 37
sediment, *35*, 38, *38*
sedimentary rocks, 9, *35*, 38–41, *39–41*
selenite, 24, *25*
shale, 45
shapes, 12–13
shells, 38
silt, 39
silver, 27, *27*
sketches, 48
slate, 42–3, *43*
smithsonite, 16, *16*
smoky quartz, 9, 21, *21*
solar panels, 29
spodumene, 29
streak plates, 51, *51*

talc, 17, *50*
testing, 25, 50, *50*
tiger's eye, 21, *21*
topaz, *31*, *50*
tourmaline, 13, *13*
tourmaline schists, 43
transparency, 30
turquoise, *31*

vinegar, 25, 51
vitreous luster, 16
volcanic glass, 37, *37*
volcano, 37

water, 24–5, 34
watermelon tourmalines, 13, *13*
waxy luster, 16
weathering, 19, 34, *35*
websites about geology, 53
wind turbines, 28, *28*

Photo Credits

iStock/rep0rter: cover, 13 (bottom).
Shutterstock/AlanMorris: 41 (top).
Shutterstock/Albert Russ: cover, 12, 20 (top), 27 (top).
Shutterstock/Aleksandr Pobedimskiy: cover, 42.
Shutterstock/Alysson M: 16 (bottom).
Shutterstock/Andriy Kananovych: 15 (bottom), 50.
Shutterstock/Anneka: 17 (top).
Shutterstock/Baloncici: 31 (second row, pearl).
Shutterstock/Bjoern Wylezich: cover, 19 (top), 20 (bottom), 23 (top), 29 (bottom).
Shutterstock/Breck P. Kent: 45 (bottom).
Shutterstock/Byjeng: 31 (second row, emerald and ruby).
Shutterstock/Cagla Acikgoz: 6–7, 9 (top).
Shutterstock/Elizabeth_0102: 31 (third row, topaz).
Shutterstock/Eniko Balogh: 30 (right).
Shutterstock/Finesell: 31 (first row, amethyst).
Shutterstock/Florian Nimsdorf: 35 (volcano inset in diagram).
Shutterstock/Honourr: 8.
Shutterstock/Inkiart: 19 (bottom).
Shutterstock/Ja Het: 46–47.
Shutterstock/James Jiao: 48.
Shutterstock/Kavic.C: 35 (diagram and pebble inset in diagram).
Shutterstock/Lea Rae: 22 (top).
Shutterstock/Leela Mei: 13 (top).
Shutterstock/losmandarinas: cover, 23 (bottom right), 41 (bottom right).
Shutterstock/luisrsphoto: 49.
Shutterstock/mahirart: cover, 37 (left).
Shutterstock/Manutsawee Buapet: 31 (first row, diamond).
Shutterstock/MarcelClemens: cover, 26 (bottom).
Shutterstock/Michael LaMonica: 51 (top).
Shutterstock/milart: 45 (top).
Shutterstock/Minakryn Ruslan: cover, 25 (bottom), 27 (bottom).
Shutterstock/Mr.Navapruet Promthong: cover, 39 (left).
Shutterstock/ND700: cover, 21 (bottom).
Shutterstock/Nika Lerman: 31 (second row, peridot).
Shutterstock/Nyura: 31 (third row, turquoise), 52.
Shutterstock/olpo: cover, 14.
Shutterstock/photo33mm: 31 (first row, aquamarine).
Shutterstock/photo-world: cover, 17 (bottom).
Shutterstock/Rattachon Angmanee: cover, 43 (middle).
Shutterstock/RHJPhtotos: 15 (top), 30 (left).
Shutterstock/salajean: 38.
Shutterstock/Sebastian Janicki: cover, 18 (bottom), 20 (middle), 21 (top).
Shutterstock/Semiglass: 24.
Shutterstock/SHTRAUS DMYTRO: 41 (bottom left).
Shutterstock/sonsart: 39 (bottom right).
Shutterstock/SpotLuda: cover, 29 (top).
Shutterstock/STUDIO492: 31 (first row, garnet), 31 (second row, sapphire).
Shutterstock/TinaImages: 31 (third row, opal).
Shutterstock/Tyler Boyes: cover, 22 (bottom), 43 (bottom).
Shutterstock/Valerie2000: 40.
Shutterstock/Viktoria Prusakova: 16 (top).
Shutterstock/Vincent Lekabel: cover, 23 (bottom left).
Shutterstock/Vladimka production: 28.
Shutterstock/vvoe: cover, 10–11, 18 (top), 26 (top), 32–33.
Shutterstock/www.sandatlas.org: cover, 9 (bottom), 36 (top), 41 (bottom middle).
Shutterstock/Yes058 Montree Nanta: cover, 37 (middle), 37 (right), 39 (top right), 43 (top), 44, 51 (bottom).
Shutterstock/Zelenskaya: cover, 36 (bottom).
Wikimedia Commons/Alexander Van Driessche - CC BY 3.0 DEED: 25 (top).